YOUR AMAZING STOMACH

DWAYNE HICKS

New York

Published in 2023 by The Rosen Publishing Group, Inc.
29 East 21st Street, New York, NY 10010

First Edition

Editor: Greg Roza
Designer: Michael Flynn

Photo Credits: Cover, p.1 LightField Studios/Shutterstock.com; p. 5 Evgeny Atamanenko/Shutterstock.com; p. 7 Explode/Shutterstock.com; p. 9 VeronikAstra/Shutterstock.com; p. 11 Nicolas Primola/Shutterstock.com; p. 13 Andrea Danti/Shutterstock.com; p. 15 Magic mine/Shutterstock.com; p.17 Anatoliy Karlyuk/Shutterstock.com; p. 19 Orawan Pattarawimonchai/Shutterstock.com; p. 21 Prostock-studio/Shutterstock.com.

Cataloging-in-Publication Data

Names: Hicks, Dwayne.
Title: Your amazing stomach / Dwayne Hicks.
Description: New York : Powerkids Press, 2023. | Series: Your amazing body | Includes glossary and index.
Identifiers: ISBN 9781725339736 (pbk.) | ISBN 9781725339750 (library bound) | ISBN 9781725339743 (6pack) | ISBN 9781725339767 (ebook)
Subjects: LCSH: Stomach–Juvenile literature. | Digestive organs–Juvenile literature.
Classification: LCC QP151.H54 2023 | DDC 612.3'2–dc23

Manufactured in the United States of America

Some of the images in this book illustrate individuals who are models. The depictions do not imply actual situations or events.

CPSIA Compliance Information: Batch #CSPK23. For Further Information contact Rosen Publishing, New York, New York at 1-800-237-9932.

CONTENTS

I'm Hungry!

Have you ever wondered why your tummy growls? That's your stomach letting you know that you are hungry! Your stomach is a sac that collects the food you eat. It also helps break down food so the body can use it.

The Digestive System

Your stomach is part of your digestive **system**. These are the parts in your body that digest—or break down—food. Then, the digestive system takes **nutrients** from the broken-down food to be used throughout the body.

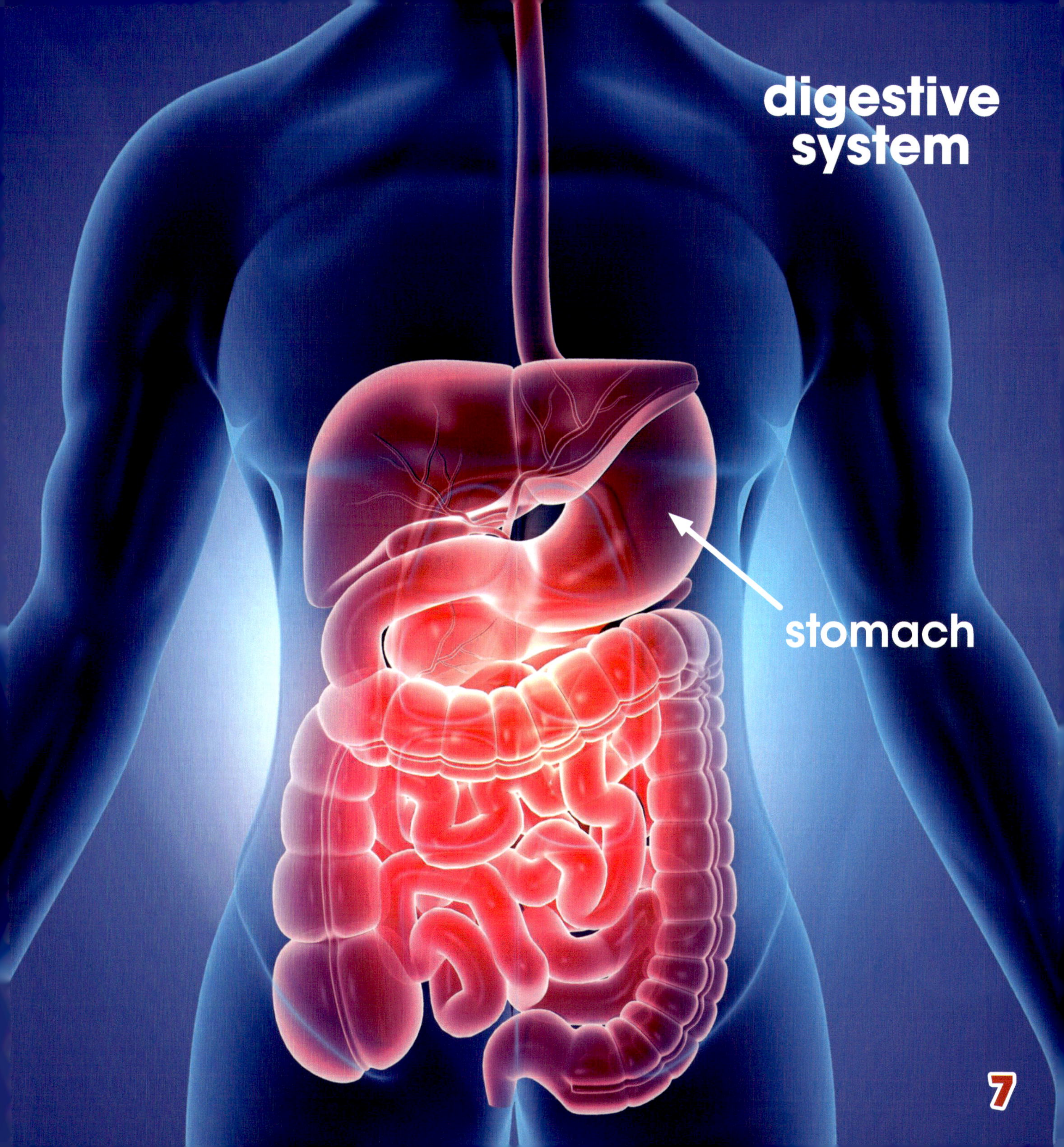
digestive
system
stomach

I'm Full!

Your stomach is smaller when it's empty. The stomach **stretches** larger when it is full of food. Your stomach's first job is to store the food you eat. Two special **muscles** keep both ends of the stomach closed.

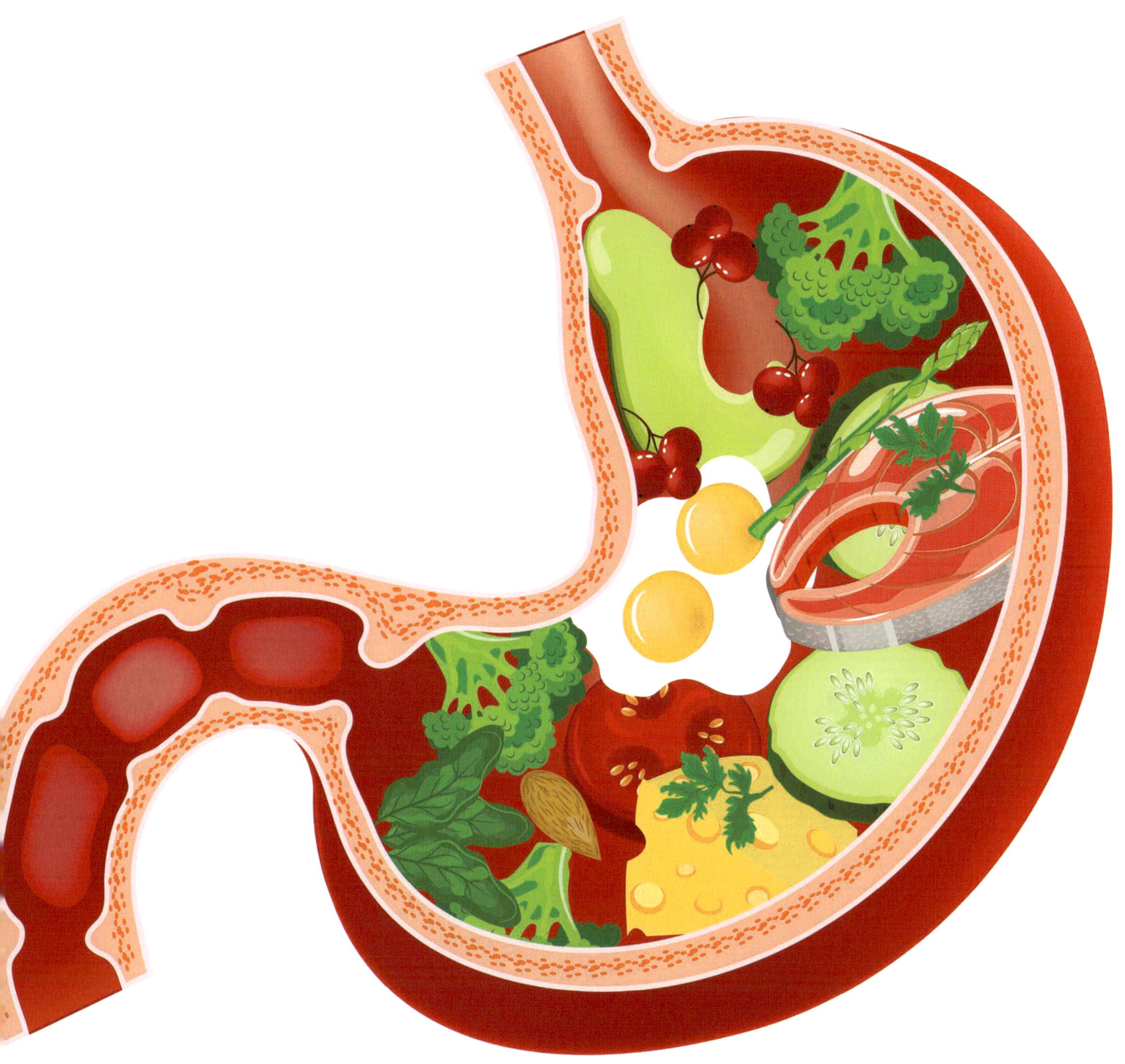

Break It Down

The stomach's second job is to digest the food you eat. The walls of your stomach have many tiny **glands** that create liquids. These liquids start breaking down food. One of the liquids is an **acid** that helps break down food and kill germs.

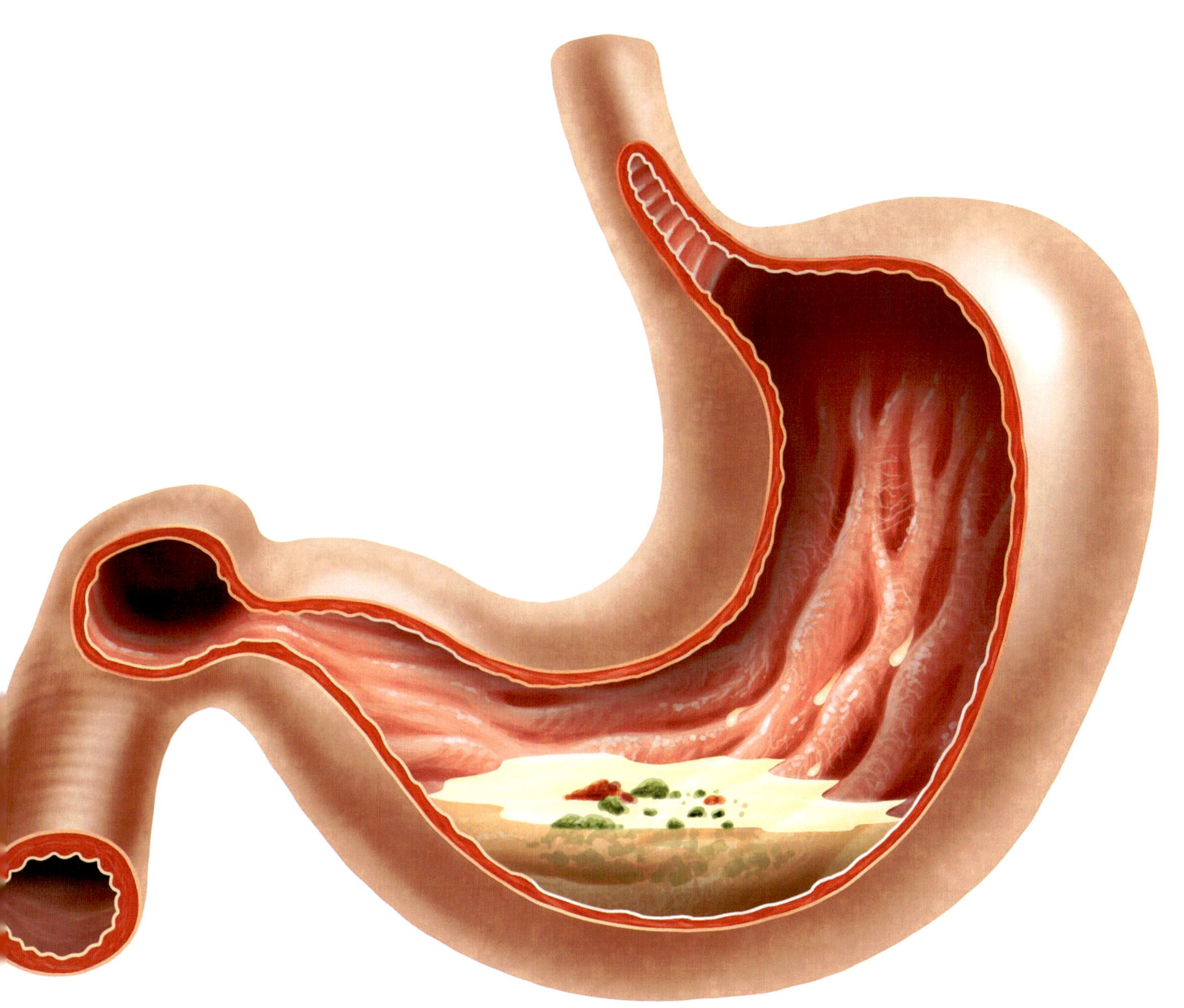

Mix It Up

The food in your stomach becomes a thick liquid. The walls of your stomach are lined with smooth muscles. This type of muscle does its job without you even knowing it! Smooth muscles mix up the digested food.

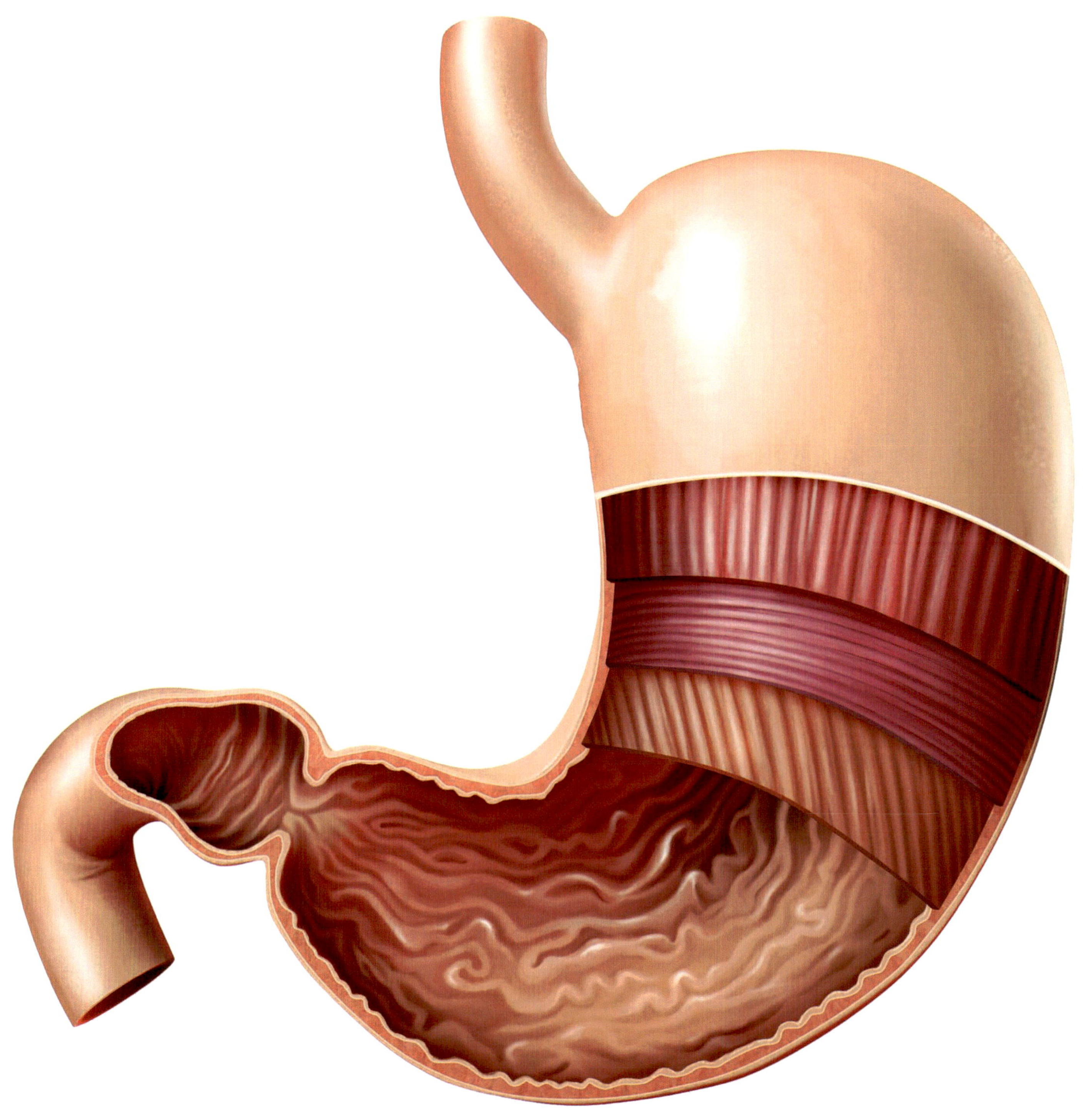

Move It Along

The stomach empties food into the small intestine. Many more nutrients are taken from the food in this body part. The smooth muscles in your stomach keep moving as it becomes empty. The muscles are now telling you that you are hungry again!

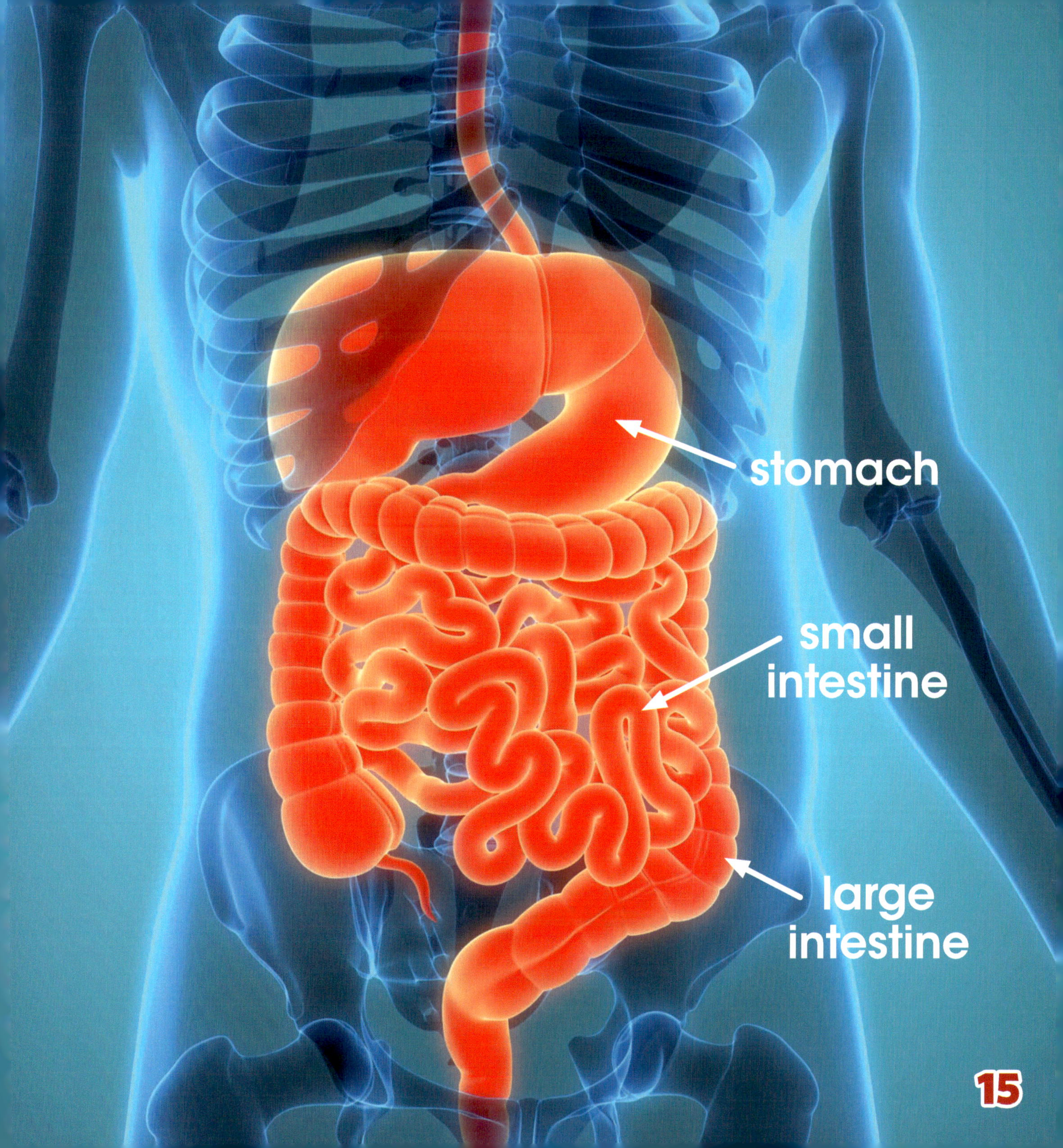
stomach
small intestine
large intestine

Buuuuurp!

You can get a stomachache from eating too much or too fast. When you eat too fast you also swallow air. This also happens when you drink sodas. What happens to that air in your stomach? It comes back out as a burp!

What Is Heartburn?

Some people get heartburn. This is when the top muscle in your stomach can't keep the food and stomach acid in the stomach. This causes a burning feeling in the throat. There are **medicines** to treat heartburn. Most lower the amount of acid the stomach makes.

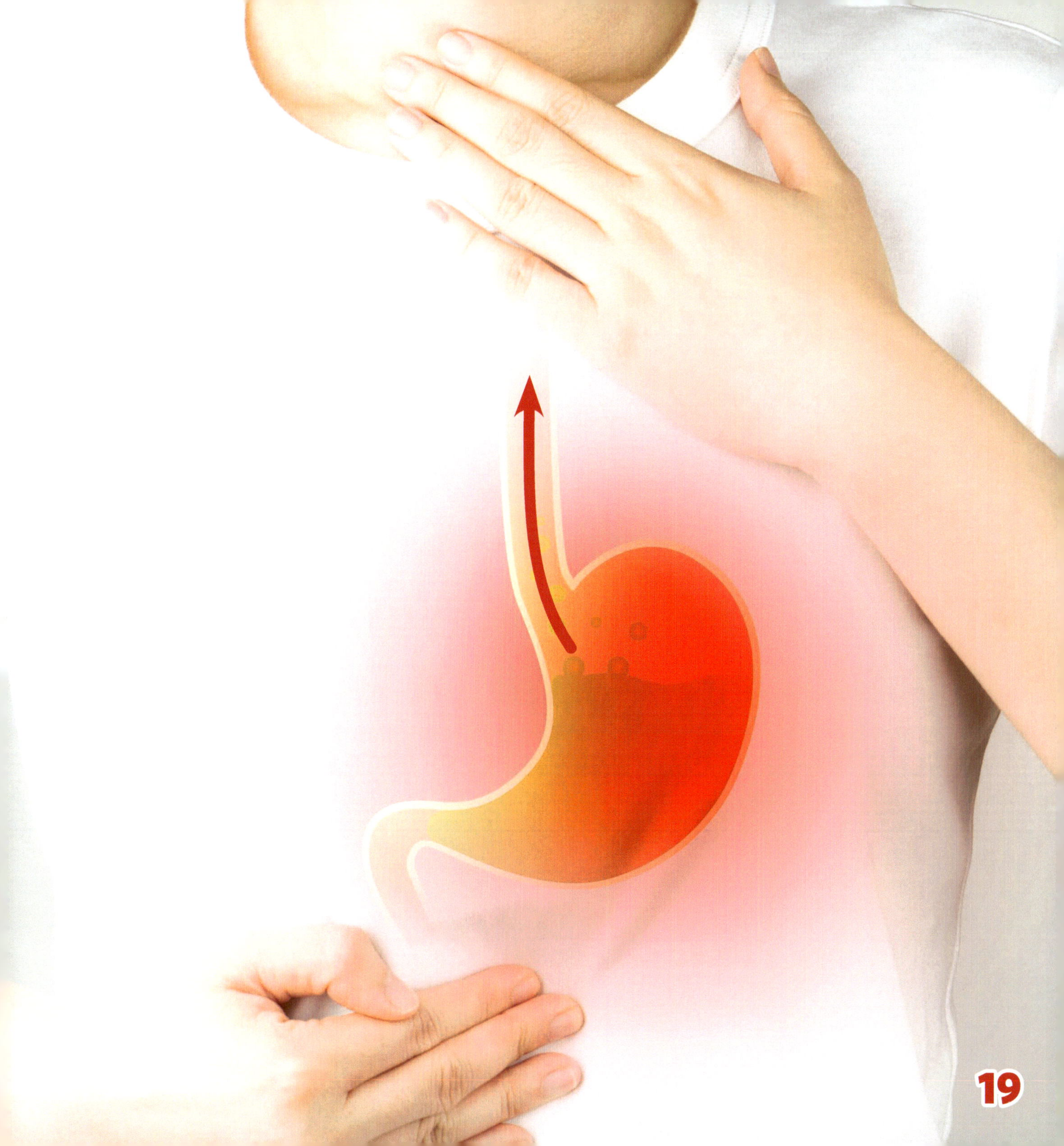

Stomach Health

You can keep your stomach and the rest of your body healthy by eating healthy foods. Foods high in fiber are good for the digestive system. Always drink plenty of water. Smoking and alcohol aren't good for the stomach and can lead to stomach illnesses.

GLOSSARY

acid: A type of liquid that can often break down matter.

gland: A part of the body that produces a substance to be used by the body.

medicine: A drug that a doctor gives you to help fight illness.

muscle: A part of the body that produces motion.

nutrient: Something taken in by a plant or animal that helps it grow and stay healthy.

stretch: To force something to grow larger without breaking.

system: A group of body parts that work together to perform an important function in the body.

FOR FURTHER INFORMATION

WEBSITES

Human Digestive System
www.ducksters.com/science/digestive_system.php
You can view a complete diagram of the digestive system at this website.

Your Digestive System
kidshealth.org/en/kids/digestive-system.html
Visit this website to learn much more about the digestive system.

BOOKS

Howell, Izzi. *The Digestive System*. London, UK: Wayland, 2018.

Ogden, Charlie. *Super Stomach and the Digestive System*. New York, NY: Booklife, 2018.

Publisher's note to parents and teachers: Our editors have reviewed the websites listed here to make sure they're suitable for students. However, websites may change frequently. Please note that students should always be supervised when they access the internet.

INDEX